What Is CHEMISTRY?

Rebecca Woodbury, Ph.D., M.Ed.

Gravitas Publications Inc.

What Is
CHEMISTRY?

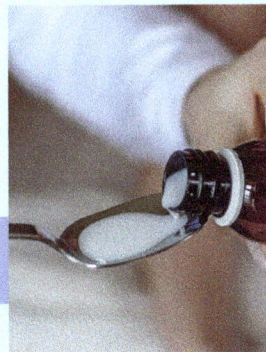

Illustrations: Janet Moneymaker

What Is Chemistry?
ISBN 978-1-950415-08-3

Published by Gravitas Publications Inc.
Imprint: Real Science-4-Kids
www.gravitaspublications.com
www.realscience4kids.com

RS4K

Image credits: Cover & Title page, Sodium and water reaction, By Alexandre, Adobe Stock; Above, By SeventyFour, AdobeStock; P. 3, By Elnur, AdobeStock; P. 13, By misskaterina, AdobeStock; P. 15, By SeventyFour, AdobeStock; P. 17, By SeventyFour, AdobeStock; P. 19, By didesign, AdobeStock; P. 21, By Sergey Nivens, AdobeStock

Chemistry is a branch of **science**.

Scientists who study chemistry are called **chemists**.

Chemists want to know what the world is made of.

Me too!

The first chemists
were called **alchemists**.

Alchemists cooked
metals, liquids, and dirt.

Alchemists did this to
make new things.

I never cook!

Greek alchemists tried
to turn lead into **gold**.

They wanted to get rich!

Could they
make cheese?

Chinese alchemists tried
to make gold very pure.

They thought pure gold
would help them live forever.

Good idea.

Chemistry happens
all around the world.

We can make
a snow mouse!

You use chemistry every day.

Yes!

Chemistry happens in your tummy when you take medicine.

Hooray for chemistry!

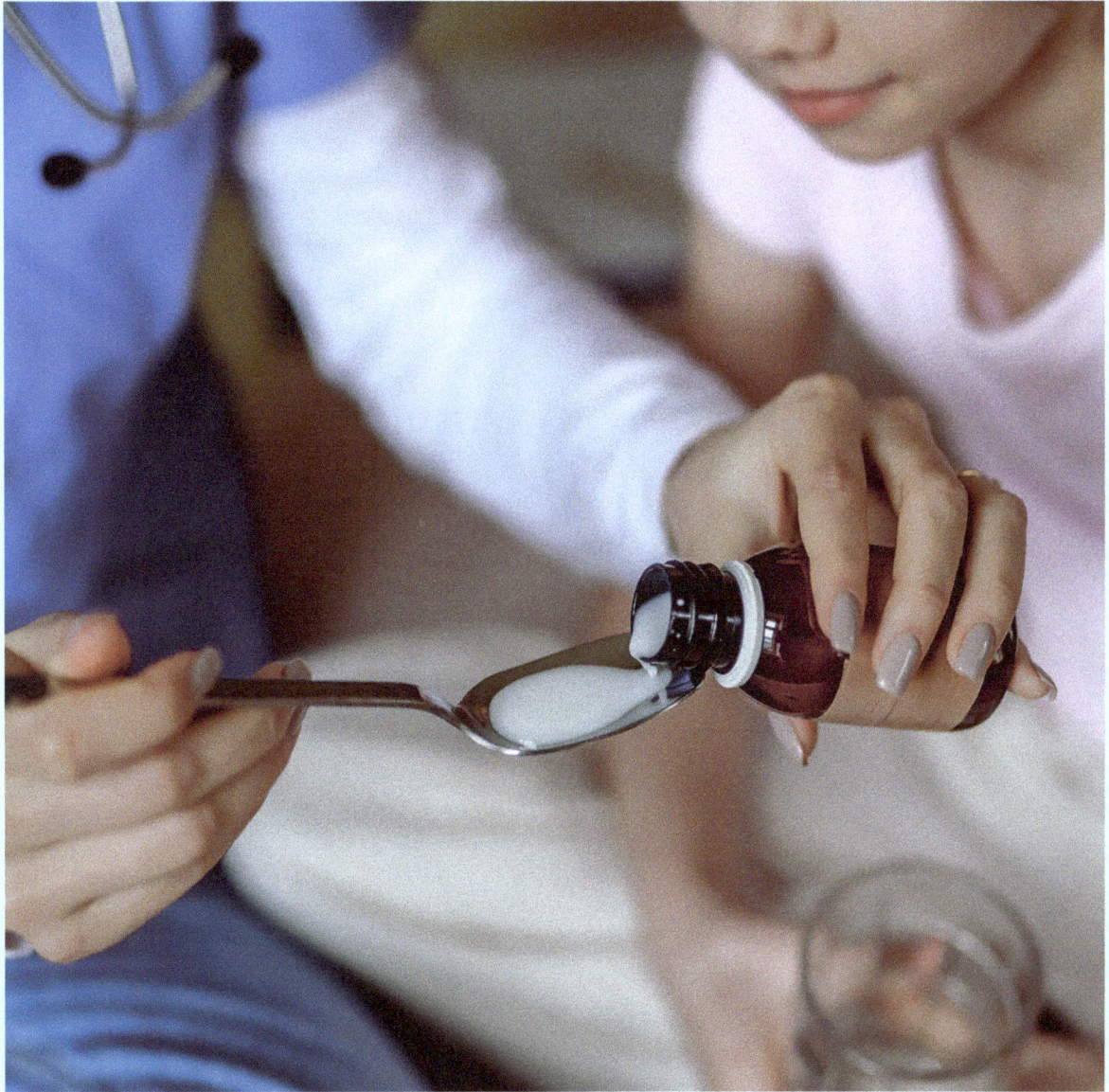

Chemistry happens when
you paint with watercolors.

Chemistry happens when
you brush your teeth.

Where did I put
my toothbrush?

I have not
seen it.

Chemistry happens everywhere.

Chemists use chemistry to help us understand the world.

I want to be a chemist!

How to say science words

alchemist   (AL-kem-ist)

chemist   (KEH-mist)

chemistry   (KEH-muh-stree)

liquid   (LIH-kwid)

medicine   (MEH-dih-suhn)

science   (SIY-uhns)

scientist   (SIY-uhn-tist)